# TEN-GALLON TOPPER

 Connect the dots from **1** to **10**.
Color the picture.

## OUT OF THIS WORLD!

 Connect the dots from **1** to **10**.
Color the picture.

# TURTLE TOWN

 Connect the dots from 1 to 10.
Color the picture.

# UNDER THE BIG TOP

Connect the dots from 1 to 10.
Color the picture.

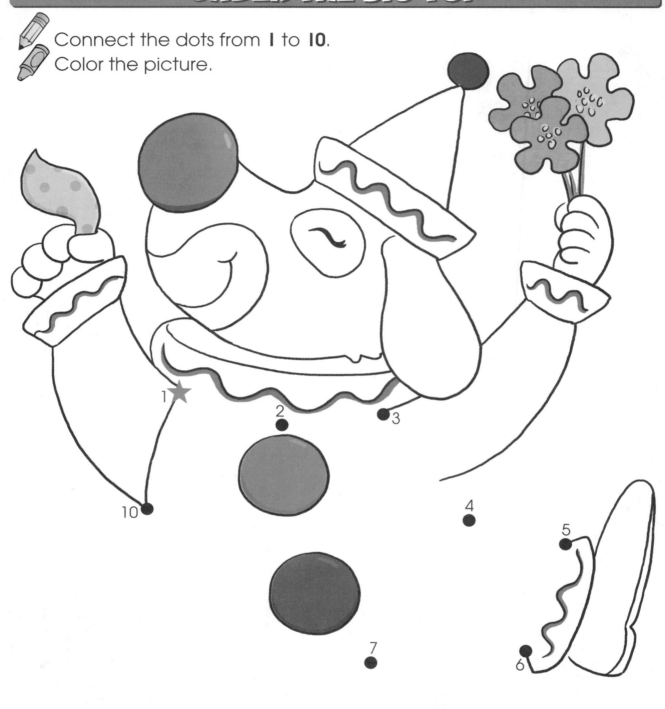

1-25 Dot-to-Dots

# WINDBLOWN

 Connect the dots from 1 to 15.
Color the picture.

# HOME-TWEET-HOME

Connect the dots from 1 to 15.
Color the picture.

## PUPPY LOVE

 Connect the dots from 1 to 15.
Color the picture.

# UNDER THE STARS

 Connect the dots from 1 to 15.
Color the picture.

# BEDTIME FOR BUNNY!

 Connect the dots from 1 to 15.
Color the picture.

# POND PALS

Connect the dots from **1** to **15**.
Color the picture.

# SAY "CHEESE!"

 Connect the dots from **1** to **15**.
Color the picture.

# FLUTTER BY

Connect the dots from 1 to 15.
Color the picture.

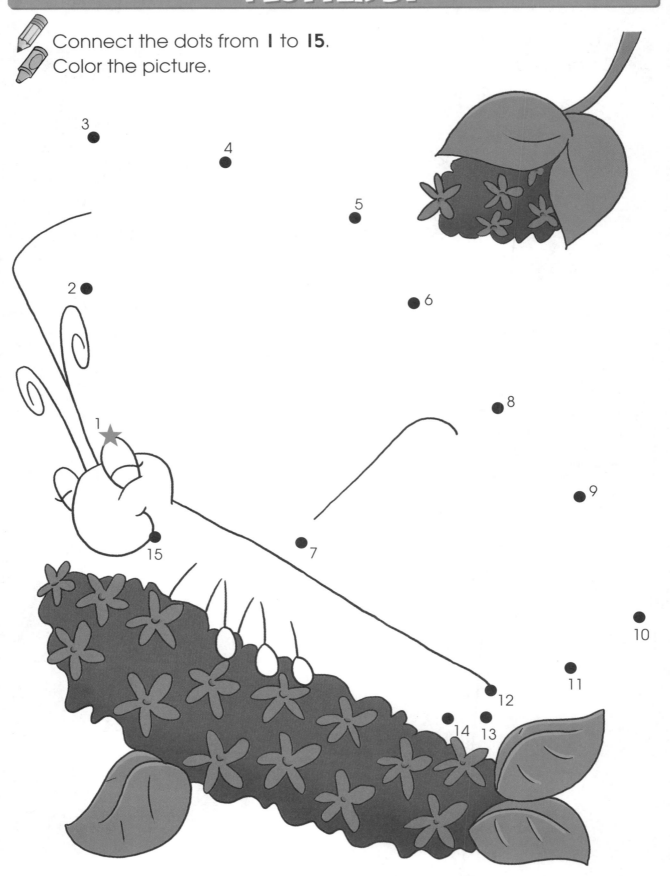

# SOMEONE'S CRABBY!

 Connect the dots from 1 to 15.
Color the picture.

# IT'S COLD OUTSIDE!

 Connect the dots from 1 to 15.
Color the picture.

1–25 Dot-to-Dots

# KARATE LESSON

 Connect the dots from **1** to **20**.
Color the picture.

# FOREST FRIEND

 Connect the dots from **1** to **20**.
Color the picture.

## GLIDING ALONG

 Connect the dots from 1 to 20.
Color the picture.

## CUTE AND COLORFUL

 Connect the dots from **1** to **20**.
Color the picture.

# AN EVENING CRUISE

Connect the dots from **1** to **20**.
Color the picture.

# HAVE YOU ANY WOOL?

 Connect the dots from **1** to **20**.
Color the picture.

# OCEAN GIANT

 Connect the dots from **1** to **20**.
Color the picture.

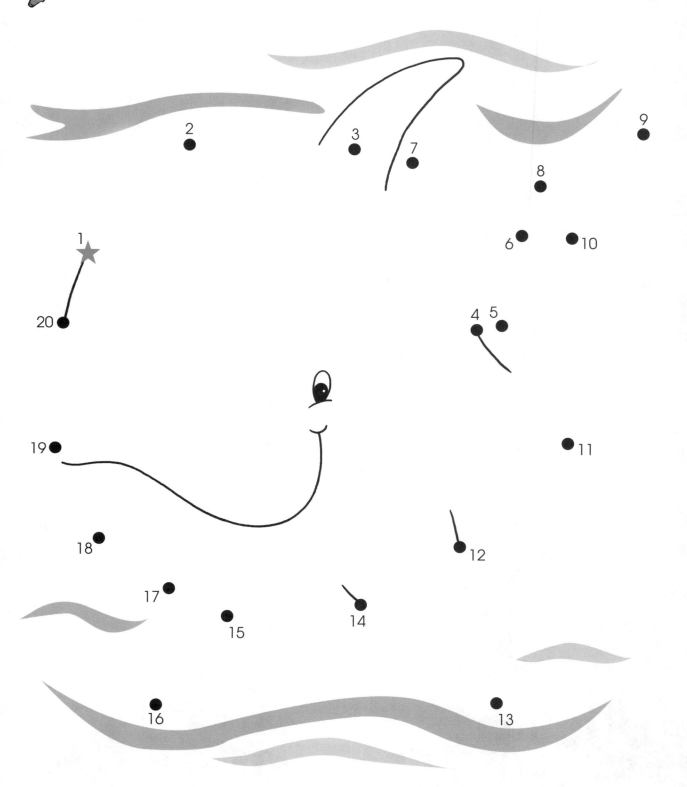

# WOODLAND BUDDIES

 Connect the dots from **1** to **20**.
Color the picture.

# OUT TO SEA

 Connect the dots from 1 to 20.
Color the picture.

# GRAZING ON THE HILL

Connect the dots from **1** to **25**.
Color the picture.

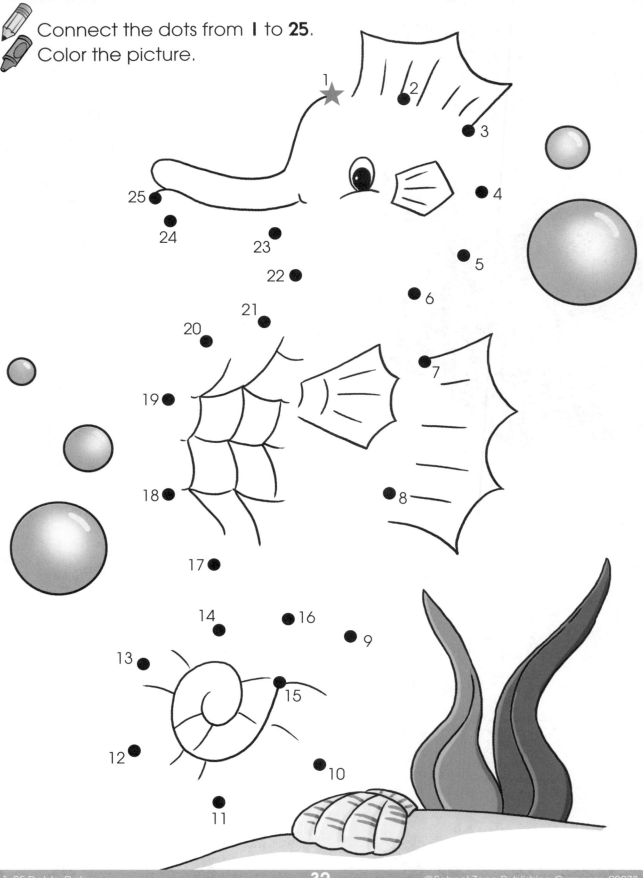

## BLENDING IN

 Connect the dots from 1 to 25.
Color the picture.

# MAKE A WISH!

 Connect the dots from 1 to 25.
Color the picture.

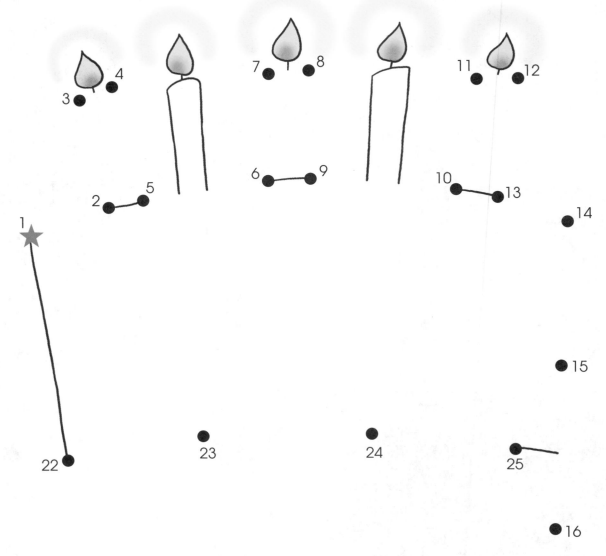

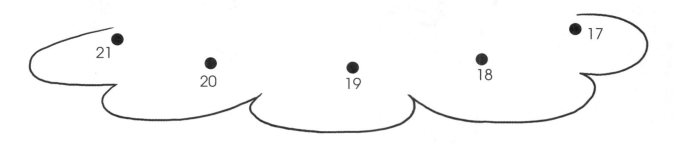

# WHAT'S COOKING?

 Connect the dots from 1 to 25.
Color the picture.

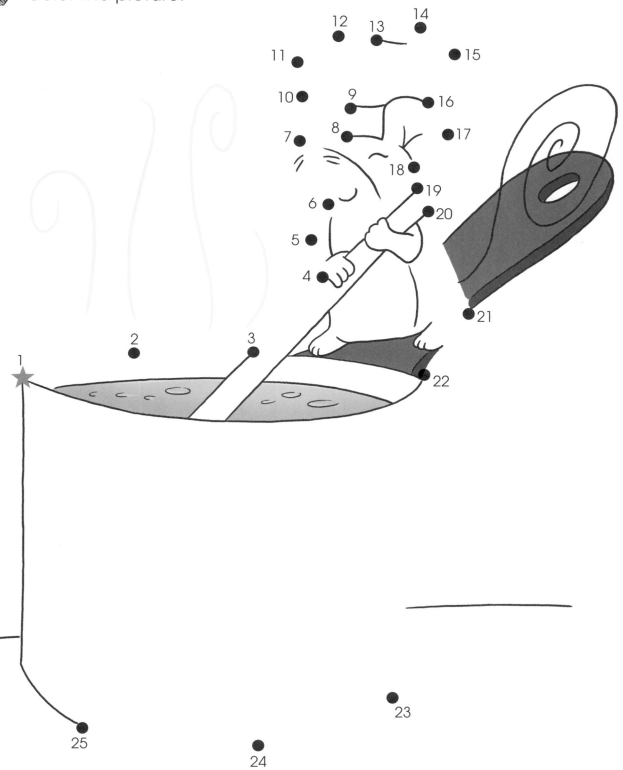

# TOUGH GUY

 Connect the dots from **1** to **25**.
Color the picture.

1–25 Dot-to-Dots

# TO THE RESCUE!

Connect the dots from **1** to **25**.
Color the picture.

## UP, UP, AND AWAY!

 Connect the dots from 1 to 25.
Color the picture.

# WHITE-WATER FUN

 Connect the dots from **1** to **25**.
Color the picture.

# DESERT DWELLER

Connect the dots from **1** to **25**.
Color the picture.

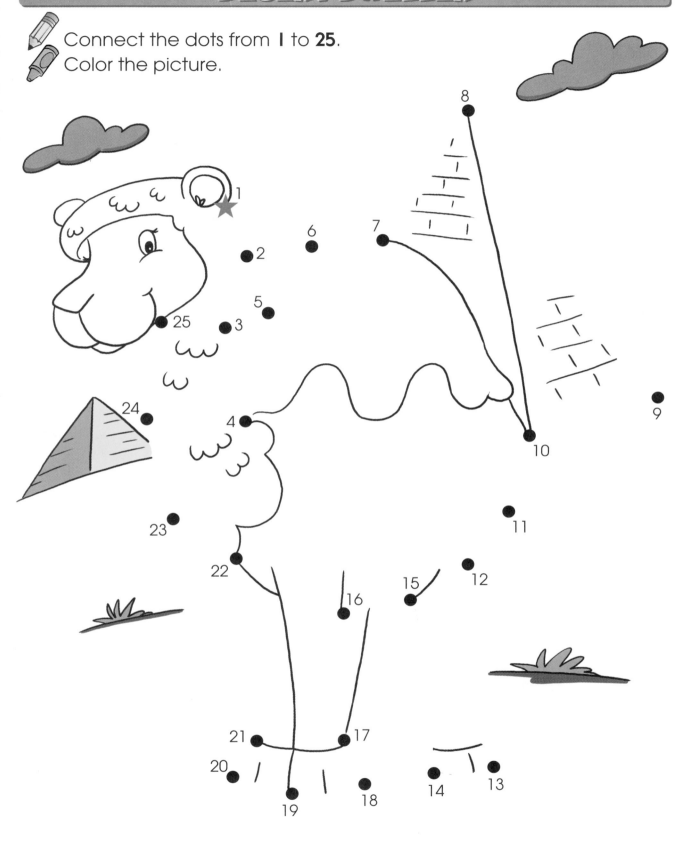

# LOST IN A GOOD BOOK

 Connect the dots from **1** to **25**.
Color the picture.

# YOO-HOO, KANGAROO!

Connect the dots from **1** to **25**.
Color the picture.

# PRACTICE MAKES PERFECT!

 Connect the dots from **1** to **25**.
Color the picture.

# OUT ON A LIMB

 Connect the dots from 1 to 25.
Color the picture.

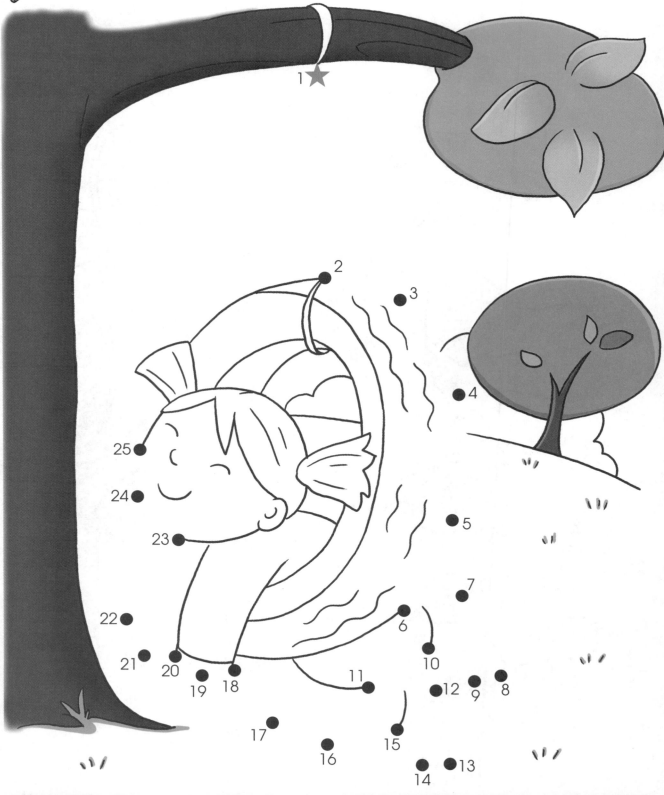

# DON'T LET GO!

 Connect the dots from 1 to 25.
Color the picture.

# OCEAN EXPLORER

Connect the dots from **1** to **25**.
Color the picture.

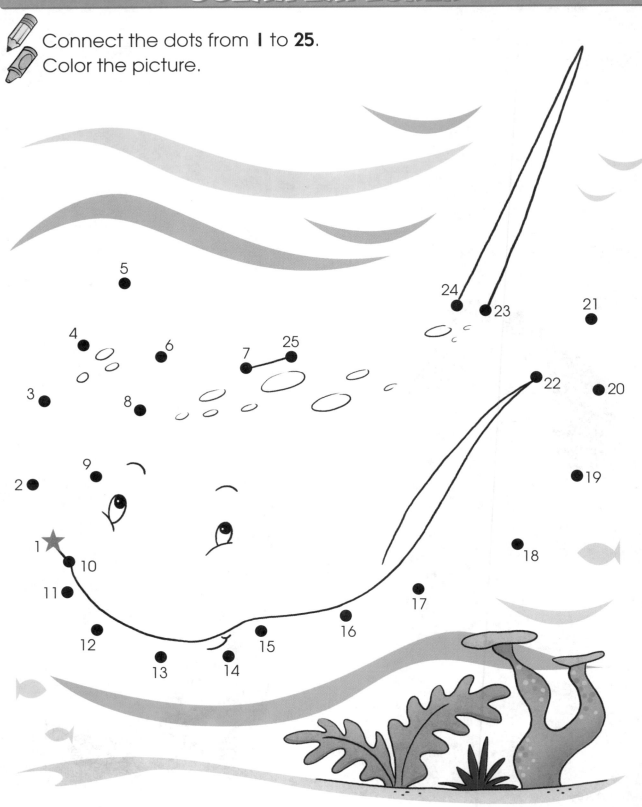

# CHATTERBOX

Connect the dots from **1** to **25**.
Color the picture.

## GOING FISHING

 Connect the dots from 1 to 25.
Color the picture.

# WHAT A HOOT!

 Connect the dots from **1** to **25**.
Color the picture.

# HAPPY TRAILS!

Connect the dots from **1** to **25**.
Color the picture.

# A LITTLE BIT COUNTRY

 Connect the dots from 1 to 25.
Color the picture.

# NIGHTTIME FLYER

 Connect the dots from 1 to 25.
Color the picture.

# BEST BUDDIES

 Connect the dots from 1 to 25.
Color the picture.

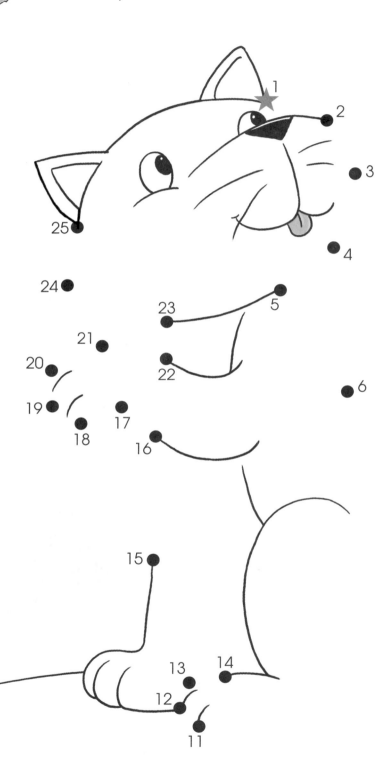
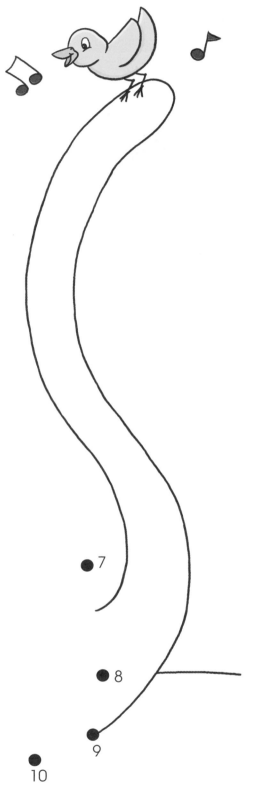

# RACE TO THE FINISH

 Connect the dots from 1 to 25.
Color the picture.

# OUT ON THE ICE

 Connect the dots from **1** to **25**.
Color the picture.

# IT'S TIME FOR THE SHOW!

 Connect the dots from **1** to **25**.
Color the picture.

# MIGHTY ROAR

Connect the dots from 1 to 25.
Color the picture.

# DINO-MITE!

 Connect the dots from **1** to **25**.
Color the picture.

# UNDER THE SEA

 Connect the dots from **1** to **25**.
Color the picture.

# BIRD'S-EYE VIEW

 Connect the dots from **1** to **25**.
Color the picture.

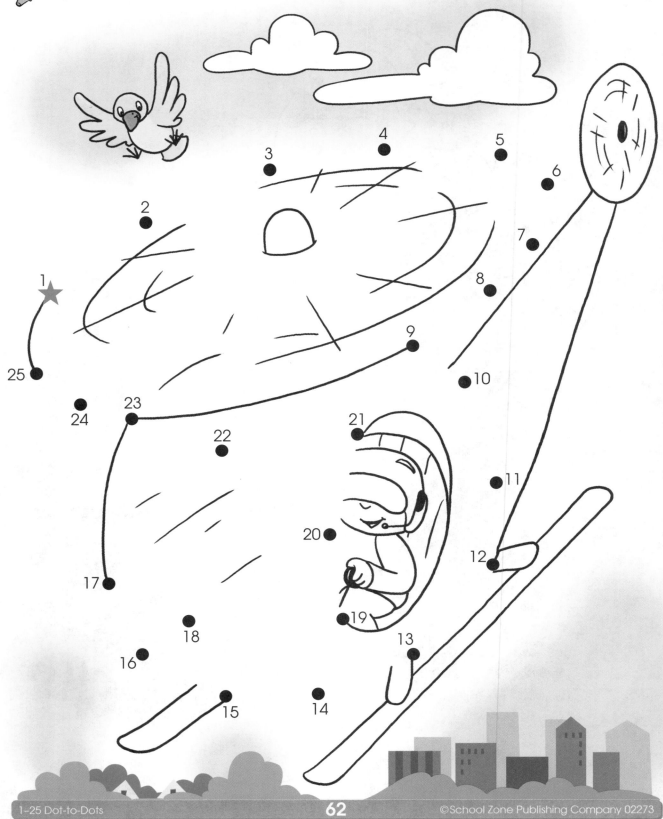

# BATTER UP!

 Connect the dots from 1 to 25.
Color the picture.